Sustainable Development & GM Food

An analysis of the relationship between the genetic modification of crops and the varieties of sustainable development

Brian Meadows

Revive Publications

*A catalogue record for this book is available
from the British Library*

ISBN: 978-1-907962-06-6

Published by Revive Publications

Reading, England

For Marco

Contents

Preface   7

Introduction   9

1   GM Food as an International Environmental Problem   11

2   The Varieties of Sustainable Development   13

3   GM Food Advances the Goal of Sustainable Development   19

4   GM Food Doesn't Advance the Goal of Sustainable Development   24

5   Summary of the Relationship between GM Food and Sustainable Development   44

6   Evaluation of the Relationship between GM Food and Sustainable Development   48

Bibliography   54

Preface

I hope that the following consideration of the relationship between sustainable development and the genetic modification of crops is of some value to you. The concepts of sustainable development and genetic modification are both wide-ranging; they can mean very different things to different people.

The issues at stake in the GM food debate clearly have a much wider significance; they tap into fundamental questions such as the nature of the relationship between humans and the rest of life on Earth, the significance of technology, and ethical issues concerning the use of technology. It is worth bearing these wider issues in mind as one considers the relationship between GM food and sustainable development.

Introduction

The relationship between genetically modified food (GM food) and the goal of sustainable development is highly complex. In order to help elucidate the relationship GM food will be framed as an environmental problem in *Chapter One*. Then, in *Chapter Two*, the various goals and definitions of sustainable development will be outlined. The arguments that GM food advances the goal of sustainable development are then considered in *Chapter Three*, and the opposing counter-arguments are considered in *Chapter Four*. A summary of the main points concerning the relationship between sustainable development and GM food is to be found in *Chapter Five*. Finally, in *Chapter Six*, an overall evaluation of the relationship is to be found.

Chapter 1

GM Food as an International Environmental Problem

The genetic modification of crops, which produces GM food, can be seen as a contemporary international environmental problem as it fulfils the seven criteria outlined by Blowers and Leroy (1996, pp. 259-261). The procedure is anthropogenic in origin, has potential global reach, is derived by experts and has differential impacts within and between countries. It also has implications for the survival of biodiversity, involves ethical arguments and gives rise to a plethora of conflicting interests.

GM food can thus be viewed as a contemporary international environmental problem. However, the question of whether it can help to achieve sustainable development is debatable and requires a definition of sustainable development.

Chapter 2

The Varieties of Sustainable Development

The goal of sustainable development inherently involves the objectives of both socio-economic well-being and environmental well-being. The obvious conflict between anthropogenic socio-economic objectives and ecocentric and anthropogenic environmental objectives gives rise to a range of definitions of sustainable development. This is because all actors will have their individual vested interests, value systems and world outlooks which give rise to differing weightings to each of the conflicting objectives.

Gibbs *et al* (1998, pp. 1351-65) outline four definitions of sustainable development. *Very weak sustainability* gives negligible importance to environmental well-being through enabling complete substitution of natural capital by human capital. *Weak sustainability* still prioritises socioeconomic well-being but does protect some natural capital. These two weak conceptions of sustainable development through prioritising development involve a piecemeal approach to environmental objectives. These approaches are favoured by business and government and are exemplified by ecological modernisation theory and extended product responsibility. *Strong sustainability* gives priority to environmental well-being through defining ecological limits which constrain socio-economic activities and protect natural capital. *Very strong sustainability* gives complete priority to environmental well-being through asserting that, rather than setting limits, all environmental resources should be strictly conserved.

The two strong conceptions of sustainability could be favoured as the goal of sustainable development on both anthropocentric and ecocentric grounds, although the latter is more likely. A further type of sustainability is purely ecocentric – the deep ecological sustainability paradigm. This can be argued to be the only goal of sustainable development which will result in genuine long-term sustainability. This is achieved by a radical transformation of socio-economic activities to bring harmony between the human world and the natural world. Fundamental principles of the paradigm are de-centralisation, bioregionalism, transport minimisation and egalitarianism (Young, 1992, pp. 14-15, 17-20).

There are thus five different – very weak, weak, strong, very strong, deep ecological – sustainable development definitions. However, each of these conceals a multitude of differing sub-definitions. For example, in the case of strong sustainability, the calculation of the appropriate ecological limits by differing agents could give rise to a whole plethora of goals rather than a singular goal. Indeed, Opschoor and Reijnders (Latesteijn and Schoonenboom, 1996, p. 237) stress that there are a whole range of sustainable development definitions due to both variations in scientific analysis and differing ethical value judgements. Nevertheless, these definitions will be used to evaluate the impacts of GM food.

In addition, further insights into the sustainability effects of GM food may be gleaned from the four sustainability paradigms of the Dutch Scientific Council for Government Policy (Latesteijn and Schoonenboom, 1996, pp. 237-238). The *utilising paradigm* involves environmental robustness and self-regulating technology. The *saving paradigm* requires reduced consumption to reach the environmental absorption capacity, due to resistance to change in economic production systems. The *managing paradigm* seeks to balance environmental and socio-economic demands through new technology, due to social resistance to lower consumption. Whilst the *preserving paradigm* necessitates a raft of societal adaptations from a flexible society to protect a vulnerable environment.

Chapter 3

GM Food Advances the Goal of Sustainable Development

A range of arguments can be made in favour of the proposition that GM food advances the goal of sustainable development. It can be argued that favourable 'input' traits lead to the reduction of pesticide and fertiliser use. This enhances environmental sustainability through both reducing resource use and the environmental damage resulting from their application. Romanian farmers are already growing potatoes using 40 per cent less insecticides than is traditionally possible (Monsanto,

2002, p. 16). The reduction of chemical inputs combined with increased yields from GM food, as has been achieved in the case of potatoes, corn and soybeans (Monsanto, 2002, p. 16) can also feed a growing human population without the need for a greater agricultural land area. This could be a great help for environmental sustainability.

Further benefits could be gleaned from the 'output' traits of GM food. Through enriching the nutritional content of food the health and productivity of millions of people in the South could be bolstered. A good example is the development of a "golden rice" that can provide Vitamin A (Monsanto, 2002, p. 16). This is another way in which GM food can enhance development in the South.

The impact of GM food on the developing countries of the South is particularly important given their need for both socio-economic development and environmental sustainability, and given the prospect of significant increases in future population levels. GM food has the potential to play a very significant role in the development of the South, through reducing chronic malnutrition, boosting productivity, increasing yields and enabling degraded and marginalised land to be utilised. It could also allow developing countries to benefit from their abundant biodiversity. However, the debates on biotechnology have been framed by the Northern countries. The Europe-US debate on allergies and toxic-effects has

21

prevented the effective utilisation of GM food technology in the South.

At a global level GM crops could play a key role in the anthropogenic adaptation process to climate change. Modification could enable crops to grow in areas where climate change has caused desertification or salinisation, areas which would be concentrated in the South. Limitation efforts could also be bolstered through fast growing plants that pull carbon out of the atmosphere, and through greater crop adaptability and yields preventing deforestation.

At a broad level GM food can be seen as the latest technological innovation. The UN 2001 Human Development Report (HDR) (p. 3) identifies three reasons why such innovation should continue. These are the potential benefits, risk management and greater costs of inertia than costs of change. Ravetz and Brown (1989, pp. 71-4) argue that regardless of these reasons it is not feasible to stop scientific and technological innovation and application. There seems little doubt in the area of GM food that in the future this technological juggernaut will provide many crops that aid the quest for sustainable development; this is in accordance with the *managing paradigm* outlined earlier.

Chapter 4

GM Food Doesn't Advance the Goal of Sustainable Development

Let us now consider the arguments which can be made that GM food does not advance the goal of sustainable development. A central argument relates to the numerous potential risks that GM crops give rise to. These risks can be separated into risks to human health and risks to the environment, as is done in the UN 2001 HDR (pp. 4, 9).

In terms of the risks to human health there is the possibility that novel genes could cause allergic responses in some people. The expression of allergenic proteins has occurred with the expression of Brazil nut protein in soy-beans (UN 2001 HDR, p. 9). A second risk is toxicity, resulting from either the introduction or increase of toxic compounds through novel genes. A third risk is pleiotropy, the occurrence of apparently unrelated multiple side-effects caused by single genes. A further risk results from the use of antibiotic markers; increased exposure to these antibiotics could result in infections becoming resistant to antibiotics.

The environmental risks concern ecosystem disruption and biodiversity loss. Evidence for unintended effects on non-target species has been found in laboratory studies, with damage occurring to monarch butterfly larvae feeding on Bt plant pollen (UN 2001 HDR, p. 9). A second risk is that the transfer of transgenic resistance traits to weedy relatives could cause greater weediness. A third risk relates to transgenic crops themselves becoming problem weeds due to their resistance traits. A fourth risk is the likelihood of the development of pest resistance to the pest-protected plants. A fifth risk is the possibility of the emergence of new viral strains in response to engineered plants with virus resistance. A further risk to biodiversity results from gene exchange to endangered wild relatives. A final risk is the use of antibiotic markers as 'triggers' which can damage micro-organisms through exposing the soil to high antibiotic concentrations

(Action Aid, 1999, p. 24). The human health risks are likely to be global in reach, whereas the environmental risks tend to be related to individual ecosystems. For example, gene flow from a GM crop is much more likely in an ecosystem where there are a greater number of wild relatives.

Further inherent risks and uncertainties resulting from the unpredictability of GM crops are highlighted by WWF International (1996, pp. 12-14, 16). They stress the uncertainties and the limits of human knowledge through the example of gene silencing. This occurs when genes inserted into plant cells appear to randomly turn themselves off. These genes may also reactivate themselves in future generations. This unpredictable phenomenon does not occur in traditional plant breeding practices, and has been observed in rice that was engineered to be resistant to the antibiotic kanamycin. WWF also use the case of 'positional effects' to highlight the unpredictability of GM crops. This occurs because the integration site in the chromosome of a novel gene cannot be determined in advance; whilst the actual integration site is of itself essential to the function and expression of the gene. Ravetz and Brown (1989, pp. 71-4) also stress the extreme uncertainties involved resulting from inadequate data, synergistic effects, thresholds and delayed impacts.

There are thus a plethora of uncertainties and risks emanating from the use of GM crops. These contribute to the concept of the 'risk society' envisaged by Ulrich Beck (Blowers & Leroy, 1996, pp. 262-265). Beck's vision of the contemporary 'risk society' is one that is characterised by individual and societal uncertainty and insecurity due to the changing nature of environmental problems. There are several aspects to the 'risk society' scenario: the emergence of 'high consequence risks' that are neither temporally nor spatially restricted in their scope; a dependence on experts who exercise control over technologies that ultimately cannot be controlled, yet who are depended upon for protection and problem identification; and, individualisation, resulting from economic insecurity, dislocation of personal life and the weakening of 'traditional' and 'modern' integrating institutions. The combination of these aspects leads to a 'negative fatalism' (Blowers, 1997, pp. 854-9) with individuals evading responsibility.

The risks of GM crops outlined above make them a prime example of a 'high consequence risk' that has the potential for global ecological disruption. Therefore, GM food contributes to the 'negative fatalism' that is a constraining factor on sustainable development. The tendency is to concentrate on the short term and to discard longer term environmental concerns. However, there is a possibility that fatalism could lead to a 'reflexive modernisation; and thus the creation of a 'New Enlightenment' (Blowers, 1997, pp. 854-9). This occurs as individuals engage in reflection and self-criticism that leads to new forms of environmental action and solidarity. However, Blowers (1997, pp. 854-9) has several reservations about this possibility. He cites the lack of pervasive individualisation and social inequality as reasons why 'reflexive modernisation' will be limited in scope. Therefore, 'negative fatalism' will still prevail and GM food will be a constraining factor on sustainable development.

In terms of the sustainable utilisation of the Earth's resources it is far from clear that GM food is a step in the direction of sustainable development. GM food makes us more reliant on fertilisers and pesticides, when the most appropriate strategy could well be a reduction in their use. Action Aid (1999,

p. 24) details the necessity of high levels of chemical dependence for GM food and the increasing toxicity required for some chemical triggers. Whilst in terms of output yields there is a lack of conclusive evidence about the effects of genetic modification. Several studies have found yield variability across crops and regions (Global Environmental Change Programme, 1999, p. 29), whilst yield sustainability in the long-term is an imponderable.

Marsden (Global Environmental Change Programme, 1999, p. 29) expects that GM food technologies will not have any effect on either food poverty, or the integration of uneven global and regional development in food production and consumption. This is because GM food is unable to affect food distribution which is a key factor in a sustainable food supply. Another factor which could engender a more sustainable food supply is a system of low technology and local control rather than high-tech GM crops controlled elsewhere. This system would also seem to be good for environmental sustainability as the spread of GM crops is concordant with a decline in agricultural biodiversity. It has been estimated that 50 years ago India grew 30,000 varieties of rice, but that there are fewer than 50 varieties expected to be grown by 2010 (Action Aid, 1999, p. 21).

In order to further assess the sustainability effects of GM crops the risk assessment framework of the UN 2001 HDR (p. 5) can be usefully utilised.

This involves not only comparing the expected harms and benefits of GM crops against the replaced technologies, but also against possible alternative technologies. This latter analysis can be used to compare the effects of a low technology and locally controlled organic system of agriculture with GM crop production. In terms of greater sustainability through lower resource use, greater biodiversity and less environmental degradation, the organic system seems far superior. Indeed, a contrast of this analysis with a comparison of GM food against the superseded agricultural system reveals that the goal of sustainability is met to a much higher degree by organic farming than either GM crop production or traditional modern farming.

A further argument that GM crops do not advance the goal of sustainable development rests on ethical grounds. It can be argued that given the plethora of uncertainties inherent to GM crops and the likely severe effects on biodiversity that their development is not ethical. This can be argued in terms of both the considerations of future generations and from an ecocentric perspective. For example, AstraZeneca's chemical switch system can change a crop's flowering time. This would lead to a diminished food supply for insects and thus a lower population, this would then cause knock-on effects throughout the food web (Action Aid, 1999, p. 24).

Adam (Global Environmental Change Programme, 1999, pp. 29-30) identifies further 'unethical' and 'dangerous' meddling with GM crops though the compression of time. GM crops are able to engender instantaneous reproductive change, rather than this occurring between generations. They can control growth and decay for just-in-time production, and also speed up ripening and maturing processes. These interferences are not acceptable from a deep ecological sustainable development paradigm. They are also potentially dangerous. Genetic engineering of a sense has occurred for millions of years through co-evolution and trial and error. The immense speeding up of these processes could lead to severe and unforeseen consequences. It can therefore be argued to be ethically wrong from both an anthropocentric and an ecocentric sustainability paradigm.

An additional group of arguments that GM crops do not advance the goal of sustainable development relates to their effects on development in the South. Many developing countries lack the institutional and regulatory capacity to manage the risks of GM crops. Despite the avoidance of first-mover risks these countries may suffer from even greater uneven development *vis-a-vis* the North due to the emergence of GM crops.

There are several particular challenges that the developing countries face in the area of GM crops, these inhibit change and increase both risks and costs. A big problem is a shortage of skilled personnel to create the regulatory system. Another serious constraint is a lack of resources to maintain the regulatory framework. A further problem is a weak communications framework compounded by a lack of public empowerment and high illiteracy rates. A final problem is inadequate feedback mechanisms which are likely to result in inferior implementation and thus lower benefits from the technology (UN 2001 HDR, pp. 10-11). These factors restrict the potential for GM crops to enhance development in the South.

In order for GM crops to have a greater developmental impact in the South the developing countries require appropriate national policies and global support. The first national policy should be to develop and harmonise health and environmental standards across countries. Coordinated research and regulation is also necessary. The development of adaptive and possibly also applied research capacities is also required. Regulatory institutions need strengthening, as has been successfully achieved in Argentina and Egypt. The views of local people, farmers and consumers also need to be sought. At a global lever there needs to be an emphasis on more research and research on longer time scales. Public trust in science needs restoring and there needs to be greater donor assistance for capacity building. Also, there needs to be greater sharing of information, as is possible with the biosafety clearinghouse of the Cartagena Protocol on Biosafety (UN 2001 HDR, pp. 11-15).

However, the main problem for the South is that the controllers of genetic engineering programmes in the North have focused their efforts on GM crop developments for the North rather than the South. These programmes are heavily involved in finding alternatives to primary products that are currently imported from the South. Examples of such programmes are the production of lauric oils, quinine, cocoa butter, sweeteners and pyrethrum (WWF International, 1996, pp. 12-14, 16). It thus seems likely that GM crops could be another tool for the North to continue exploiting the South, rather than a tool that could aid development. It is thus not surprising that the OECD's analysis of biotechnology concludes that: "there is strong evidence that developing countries, notably those heavily engaged in agriculture, will bear the brunt of trade impacts for a long time to come" (WWF International, 1996, pp. 12-14, 16).

Action Aid (1999, pp. 17, 26) has highlighted two more factors that could have a negative impact on development in the South. The increased herbicide use associated with GM crops could exacerbate rural poverty and unemployment through reducing the demand for labour-intensive agricultural work. Whilst the increasing corporate patenting and promotion of GM crops in the South is likely to cause a deterioration in the livelihoods of the

41

poorest farmers; this being the result of the high costs of growing GM crops and the effect of the patents in increasing the prices of non-GM seeds.

These latter arguments for the proposition that GM food does not advance the goal of sustainable development depend more on the role of large companies in the control of the GM process than on the GM food itself. Indeed, Scoones and Newell (Global Environmental Change Programme, 1999, p. 33) argue that the concentration of biotechnology R&D expertise in a few private corporations means that it is unlikely that GM food will have a beneficial impact on poverty. This begs the question as to whether a more favourable sustainable development outcome for GM food would be achieved if the technology were not in the hands of big business. This seems very likely and has prompted calls for greater international investment in international food research stations, and universities, to develop GM crops for poor farmers (Global Environmental Change Programme, 1999, p. 29).

Chapter 5

Summary of the Relationship between GM Food and Sustainable Development

It has been argued that there are a multitude of definitions of sustainable development. Five broad definitions have been outlined – very weak, weak, strong, very strong, and deep ecological. Each gives a varying emphasis to anthropocentric versus ecocentric objectives, and this reflects a different socio-economic scenario. This variation is a result of varying ethical and scientific judgements and is reflected in the four sustainability paradigms of the

Dutch Scientific Council for Government Policy. It is in this context that the arguments for and against *the proposition that GM food advances the goal of sustainable development* have been made.

Arguments made in support of the proposition have included favourable 'input' traits, higher yields and enhanced 'output' traits. The potential benefits for development in the South have been highlighted along with the possibilities for limitation and adaptation to global anthropogenic climate change. GM food has also been framed in terms of the 'technological progress' discourse; from this perspective future benefits in terms of sustainable development seem assured.

The arguments made against the proposition have centred on a few key areas. The uncertainties and health and environmental risks inherent in GM crops make them a prime example of Beck's 'high consequence risks'. The dependence on fertilisers and pesticides, yield uncertainty, declining agricultural biodiversity and the lack of food poverty alleviation have been identified as important aspects of GM crop use. When these aspects, and the aspects of traditional modern farming, were compared to organic farming, it was clear that organic farming was superior in terms of sustainability. Ethical arguments have been made on the basis of both green values and inherent dangers. It has also been argued that GM crops will enhance uneven North-

South development, although this may partly be due to the control of GM food by big business rather than to the technology itself.

Chapter 6

Evaluation of the Relationship between GM Food and Sustainable Development

An evaluation of the extent to which GM food advances the goal of sustainable development involves contrasting these opposing arguments with the various definitions of sustainable definition. I personally believe in the deep ecological sustainable development paradigm. I believe this to be the purest type of sustainable development and the only type that can be sustained in the long-term. From this perspective and at a fundamental level I thus do not believe that GM food is in any way conducive to sustainable development. For me, sustainable development requires a critical approach rather than a problem-solving approach (Elliott, 1998, pp. 243-8). Whilst not opposed to all technological progress I believe there to be an ethical line that should not be crossed, and which if it is crossed will entail serious unforeseen and unpredictable consequences.

Whilst being sceptical of the truth and value of some of the arguments for GM food such as greater yields and improved 'input' and 'output' traits I would accept the possibility that they have the potential to achieve some development objectives in the South and to play a role in the climate change arena. This would perhaps partially depend on wresting control away from big business. However, I believe that these very same objectives could be met more fully through a local knowledge inspired organic agriculture and bioregionalism.

Proponents of weak sustainability approaches are likely to consider GM food as fulfilling sustainable development to a high extent through enabling 'win-win' outcomes. For example, greater yields on the one hand enable less deforestation on the other; and, enhanced 'input' traits on the one hand enable lower resource use on the other. However, how can this result in any kind of sustainable development when overall environmental impact levels are ignored?

Proponents of the holistic approach of strong sustainability will be more concerned about the risks and uncertainties of GM food and whether these may cause a crucial ecological limit to be breached. For them the extent to which GM food fulfils sustainable development depends on weighing up the whole range of positive and negative impacts and how they relate to pre-defined ecological limits. Some impacts could be viewed negatively *vis-à-vis* the limit such as increased biodiversity loss from gene exchange to endangered wild relatives, whilst others, such as lower pesticide use, could be viewed positively. The advocates of very strong sustainability are likely to view the risks to biodiversity inherent to GM food as not concordant with sustainable development. There are thus many opinions as to the extent to which GM food advances the goal of sustainable development, of which mine is only one.

Bibliography

Action Aid (1999) 'AstraZeneca and its Genetic Research: Feeding the World or Fuelling Hunger?', *Perspectives on Biotechnology*, Milton Keynes, The Open University.

Blowers, A. (1997) 'Environmental Policy: ecological modernisation or the Risk Society?', *Urban Studies*, vol. 34, number 5-6, pp. 845-71.

Blowers, A. and Leroy, P. (1996) 'Environment and Society: shaping the future' in Blowers, A. and Glasbergen, P. (eds.), *Environmental Policy in an International Context: Prospects,* London, Arnold.

Elliott, L. (1998) *The Global Politics of the Environment,* London, Macmillan.

Gibbs, D. C., Longhurst, J. and Braithwaite, C. (1998) 'Struggling with "sustainability": weak and strong interpretations of sustainable development within local authority policy', *Environment and Planning A,* vol.30, no. 8, August, pp. 1351-66.

Global Environmental Change Programme (1999) 'The Politics of GM Food: Risk, Science and Public Trust', *Perspectives on Biotechnology,* Milton Keynes, The Open University.

Latesteijn, H. and Schoonenboom, J. (1996) 'Policy scenarios for sustainable development' in Blowers, A. and Glasbergen, P. (eds.), *Environmental Policy in an International Context: Prospects,* London, Arnold.

Monsanto (2002) 'Biotech Primer', *Perspectives on Biotechnology,* Milton Keynes, The Open University.

Ravetz, J. and Brown, J. (1989) 'Biotechnology: anticipatory risk management' in Brown, J. (ed.), *Environmental Threats: Perception, Analysis and Management,* London, Belhaven Press.

UN 2001 Human Development Report (HDR) 'Making new technologies work for human development', *Perspectives on Biotechnology,* Milton Keynes, The Open University.

WWF International (1995) *Genetic Engineering: Examples of Ecological Effects and Inherent Uncertainties,* Gland, Switzerland, WWF International.

Young, S.C. (1992) 'The different dimensions of green politics', *Environmental Politics,* vol.1, no. 1.

www.ingramcontent.com/pod-product-compliance
Lightning Source LLC
Chambersburg PA
CBHW050539210326
41520CB00012B/2642